Libro de resúmenes de la II Jornada en técnicas de Biotecnología y Ciencias Ómicas

Editado y coordinado por el proyecto:

PIE15-110

Editores:
Rafael A. Cañas
Fernando N. de la Torre
Mª Belén Pascual
Jorge El-Azaz
Francisco R. Cantón
Francisco M. Cánovas
Concepción Ávila

Trabajo desarrollado en el marco del Proyecto de Innovación Educativa de la Universidad de Málaga
PIE15-110

ISBN: 1548111805
ISBN-13: 978-1548111809

ÍNDICE DE CONTENIDOS

Prefacio   i

Programa de las Jornadas   1

Sesión 1 Biología Molecular y Biotecnología   3

Two-hybrid. Split ubiquitin system   5

CRISPR-Cas9 as tool for lung cancer treatment   6

Advances in sequencing methods along the time   7

Synthetic DNA with capacity to evolve   8

Sesión 2 Genómica y Transcriptómica   9

Chromosome conformation capture assays   11

Chromatin immunoprecipitation   12

MiRNA bank: miRNA isolation and its clinical applications   14

Sesión 3 Proteómica y Metabolómica   **15**

HiQuant   17

Matrix Assisted Laser Desorption/Ionization-Time of Flight (MALDI-   18
TOF) Mass Spectrometry

Sesión 4 Bioinformática   **19**

Nuclear magnetic resonance, the path towards a recreation of protein   21
structures in solution

Co-expression networks in plants   22

PREFACIO

En los últimos años han tenido un gran auge las Ciencias Ómicas. Por sus características particulares éstas se han desarrollado como ciencias interdisciplinares que utilizan técnicas de laboratorio propias de la Biología o la Química y herramientas de la Bioinformática. La velocidad con la que se han desarrollado las Ciencias Ómicas ha provocado que buena parte de los planes de estudios de las carreras relacionadas con la Biología hayan quedado desfasados respecto a ciertas aproximaciones experimentales. Desde este punto de vista, se hace necesaria una aproximación de los alumnos de estas titulaciones a los fundamentos y técnicas de las disciplinas que les resultarán complementarias para el entendimiento de las Ciencias Ómicas y una futura aplicación de sus metodologías.

Esta "II Jornada en técnicas de Biotecnología y Ciencias Ómicas" se enmarca dentro del Proyecto de Innovación Educativa PIE15-110 concedido por la Universidad de Málaga. La jornada tiene como objetivo fundamental el acercamiento de los alumnos de las titulaciones de Grado en Biología y Grado en Ingeniería de la Salud. El objeto de ello es la mejora del aprendizaje y la generación de las capacidades necesarias para facilitar un futuro trabajo en Ciencias Ómicas por parte de los nuevos graduados en Biología e Ingeniería de la Salud.

En este libro se encuentran los resúmenes presentados por los alumnos participantes en el proyecto PIE15-110 en el marco de la "II Jornada en técnicas de Biotecnología y Ciencias Ómicas" que tuvo lugar en el Salón de Grados de la Facultad de Ciencias de la Universidad de Málaga el 25 de mayo de 2017. Por tanto, la autoría de cada resumen en este libro corresponde a los alumnos que los presentaron a la jornada. La organización de la jornada fue llevada a cabo por los investigadores del proyecto PIE15-110 y editores del presente libro de resúmenes:

Rafael A. Cañas

Fernando N. de la Torre

Mª Belén Pascual

Jorge El-Azaz

Francisco R. Cantón

Francisco M. Cánovas

Concepción Ávila

Programa
de la
II Jornada en técnicas de Biotecnología y Ciencias Ómicas

Celebrada en el Salón de Actos de la Facultad de Ciencias de la Universidad de Málaga el jueves 25 de mayo de 2017.

15:30h-15:40h Apertura de la Jornada por **Eva Millán Valldeperas**, Subdirectora de Ordenación Académica de la Escuela Técnica Superior de Ingeniería Informática, y **Antonio Flores Moya**, Decano de la Facultad de Ciencias.

15:40h-16:10h Conferencia invitada. *De los rasgos poligénicos a los poligenómicos.* **Prof. Manuel Gonzalo Claros.**

16:10h-16:40h Conferencia invitada. *Diseño de circuitos para Biología Sintética.* **Dr. Raúl Montañez Martínez.**

16:40h-16:55h Comunicación oral. **Sesión 1. Biología Molecular y Biotecnología.** *Two hybrid. Split ubiquitin system.* (**Lucía Llamas; Isabel Ojeda; Alejandro Nieto**).

16:55h-17:10h Comunicación oral. **Sesión 2. Genómica y Transcriptómica.** *Chromosome conformation capture assays.* (**Mónica Bustos; Andros Mahiques; Almudena Peláez**).

17:10h-17:25h Comunicación oral. **Sesión 3. Proteómica y Metabolómica.** *HiQuant.* (**Ana Ángel; Francisco Díez de Los Ríos; Miguel Ángel Gallardo**).

17:25h-17:40h Comunicación oral. **Sesión 4. Bioinformática.** *Nuclear magnetic resonance, the path towards a recreation of protein structures in solution.* (**Pablo Rodríguez; José Córdoba; José Miguel Valderrama**).

17:40h-18:00h **Entrega del premio** a la mejor presentación y al mejor póster.

18:00h-19:00h **Sesión de paneles.** Traslado al hall de Químicas.

19:00h-19:10h **Entrega del premio** al mejor póster.

Sesión 1

Biología Molecular y Biotecnología

Comunicación oral

Two-hybrid. Split ubiquitin system

Lucía Llamas; Isabel Ojeda and Alejandro Nieto

Fundamentos de Biotecnología Molecular

Yeast two hybrid is a molecular technique widely used today because it allows studying protein - protein interactions, through the direct transcriptional activation of a reporter gene in yeast. The "Split ubiquitin system" is a variant of this technique that allows the study of membrane proteins interaction. The localization of such proteins does not allow their study with the standard double hybrid, as they are not able to access the nucleus and activate transcription.

This technique consists in studying the interactions of membrane proteins: both of a protein against a collection of proteins coded in a library, as well as to see if two concrete proteins are able to interact. The difference between the double hybrid and this variant is that in this case an ubiquitin protein is divided into the N-terminal (Nub) and C-terminal (Cub), whose coding sequence are located in a different yeast expression vector. Together with the end Cub is added a transcription factor (TF). There are two types of vectors: on the one hand, a vector in which the study protein is fused to the Cub end and the transcription factor and on the other hand, a vector in which a cDNA library is cloned fused to the Nub end (with a point mutation, I13G), both vectors are transformed into yeasts. With regard to cloning it is important to bear in mind that we want a single protein to be formed and therefore stop codons must be eliminated.

The interaction between proteins will lead to the reconstruction of ubiquitin which allows the recognition by an ubiquitin-specific endogenous protease, resulting in a proteolytic cleavage and the release of the transcription factor. This transcription factor is able to enter the nucleus and activate the transcription of a reporter gene (Ade2, His3, and LacZ).

Recent use of these technologies is a step forward in the analysis of how membrane proteins interact in a cell. It can clarify signal transduction, ion channels, mechanisms leading to neurodegeneration, and interactions between viral and host proteins, where it has been identified numerous interactions of membrane protein that are crucial to cellular regulation. This technique has been used to study, for example: potassium channels in plants and the protein presenilin in Alzheimer's disease.

Bibliography:

- Jones et al. (2014). Science, 344: 711-716.

- Snider et al. (2010) J Vis Exp, 36: e1698-e1698.

- Stagljar and Fields (2002) Trends Biochem Sci, 27: 559-563.

- Cervantes et al. (2001) FEBS Lett, 505: 81-86.

Panel

CRISPR-Cas9 as tool for lung cancer treatment

Juan Jesús Rosado Cabral[1]; Paola Calvente Puertas[2] and Ana Sofía Rodríguez Mariscal[2]

[1]Genómica Estructural y Funcional; [2]Fundamentos de Biotecnología Molecular

Explaining the functionality of the genetic elements involved in certain diseases requires precise genome editing technologies. CRISPR-Cas is a gene-editing system that was discovered from genomes of bacteria and archaea, serving as a defense against viruses and plasmids. This system consists of a series of gene sequences, which encode nucleases that cut and degrade exogenous DNA at certain sites.

There are different CRISPR-Cas systems, Type I, Cas6e / Cas6f cut into the RNA junction of the single strand and double strand RNA. Type II uses a trans-activator RNA (tracr) to form double- stranded RNA, which is cut by Cas9 and RNaseIII. Types II and III, mature RNAs are produced. This mature RNAs associate with proteins to form interference complexes. In type I and II systems, the mating of bases between the RNAcr and a sequence causes degradation of the invading DNA. In type III systems do not require this sequence for degradation.

The best known is Type II, where Cas9 carries out the cut-off of the dsDNA sequence and those cuts are repaired by non-homologous recombination (NHEJ) or by direct homology recombination (HDR). NHEJ is active throughout the cell cycle and is an order of magnitude faster than HDR. However, HDR uses longer stretches of sequence homology to repair DNA lesions. Double-strand break repair systems are used for genome editing.

Using these two mechanisms, we can introduce a premature STOP, generating knock out of specific genes or even, we can introduce an exogenous gene accompanied by a DNA guide to correct a particular mutation, among many other applications, such as cancer treatment.

Using genetically modified lymphocytes by CRISPR-Cas9, scientist are trying to promote the immune system's response to remove malignant tumors. Cas9 nucleases may be targeted by short RNAs to induce a precise cleavage at endogenous genomic loci in human and mouse cells. For example, this is used as a therapeutic approach for the treatment of lung cancer where modified lymphocytes using the technique to promote the immune system response. In fact, genomic mutations in the EGFR gene were identified and it activity was restored using CRISPR-Cas 9 system.

With time, it will be introduced in clinical research for the modification in cells of the germ line, somatic or in embryos, but first it is necessary to determine its ethical limits.

Panel

Advances in sequencing methods along the time

Beatriz Aguilera[1]; María Balsera[1] and María José Muñoz[2]

[1]Genómica Estructural y Funcional; [2]Genómica, Proteómica y Metabolómica

The sequencing systems can be grouped in three distinct categories depending on their characteristics. The first one is the First-Generation Sequencing Methods with two methods. The first method, Sanger, is a very limited method, is inefficient for large-scale projects, it is also very laborious and expensive. The second method, Maxam and Gilbert have a grand problem his low resolution obtained, but it could be improved using very thins acrylamide gels.

The Second-Generation devices, also called next generation sequencing (NGS) technologies, offer the possibility to map entire genomes. Roche-454 system was the first commercially successful NGS system, it uses nebulization, emulsion PCR for amplification and pyrosequencing technology relies on the detection of pyrophosphate released during nucleotide incorporation. The sequencer Illumina adopts the technology of sequencing by synthesis. The library with fixed adaptors is denatured to single strands and grafted to the flowcell, followed by bridge amplification to form clusters which contains clonal DNA fragments. SOLID changes the amplification method from emulsion PCR to Wildfire template walking process and detect the labeled oligodNTP. Finally, the Ion Torrent system amplifies DNA by emulsion PCR and it detect if the dNTP is include by the change of the pH of the medium due to the release of a proton by the incorporation of a nucleotide.

The third generation sequencing systems are able to sequencing the DNA without amplification due to their detector is capable to detect the signal variation of a single molecule, avoiding the lag on the chain. This systems are, Oxford Nanopore, which is able to measure the characteristic disruption made by each dNTP, and Pacific Bioscience which can detect the fluorescence of a single labeled dNTP while is added to the chain by the DNA polymerase. Pacific Bioscience nowadays is able to obtain 1Gb of data, similar as the other systems, within 30 minutes to 4 hours and the length of the reads are about 12-20 Kb.

The application of these techniques are very random, it can be used in transcript expression analysis through RNA-Seq, miRNA, methylation of the DNA or polyadenylation; or in genome sequencing de novo through genome sequencing, disease gene identification, mitochondria DNA, mutation detection. Also the gene transcription controlled by transcription factors can be possible to research with this new technology. Epigenetic applications are also development as the study of the human epigenome project.

Panel

Synthetic DNA with capacity to evolve

María Evelin Quispe Molina and José Miguel Pérez Tejeiro

Fundamentos de Biotecnología Molecular

XNAs (xeno nucleic acids) consist in artificial compounds which present an altered sugar bone compared with natural nucleic acids, currently knowing six different xeno nucleic acids in total (Anosova et al, 2016). Several experiments have been made with these compounds to study their chemical and physical properties, getting to the point of producing artificial enzymes that can replicate XNAs polymers and even transform them into DNA molecules, and DNA back into XNA (Pinheiro et al, 2014). More recently, synthetic biologists Philipp Holliger and Alexander Taylor, both from the University of Cambridge, managed to create XNAzymes, the XNA equivalent of a ribozyme, enzymes made of DNA or ribonucleic acid. All these knowledge have made possible the development of a new field called xenobiology (Chaput et al, 1012).

In the field of biology, these compounds have a special interest because it has been demonstrated that they can perform in vitro the same role as DNA or RNA, storing and carrying genetic information. Despite of being this information useless and invisible for genetic mechanisms presented in organisms (Pinheiro et al, 2012); and what is more shocking, XNAs polymers can experiment modifications that can change their polymers which could be understand as a capacity to evolve and change in the time. Bearing all this in mind, it is possible to find interesting applications for XNAs, such us become important tools in molecular biology (Kasahara et al, 2016), use them as potential disease-fighting agents in medicine or understand more about the possibility that different polymers would have been implicated in the origin of life (Pinheiro et al, 2012).

Bibliography:

- Anosova et al. (2014) Nucleic Acids Res., 44: 1007-1021.

- Chaput et al. (2012) Chemistry & Biology, 19: 1360-1371.

- Kasahara, et al. (2016) Molecular BioSystems., 13: 235-245.

- Pinheiro et al. (2014) Curr Protoc Nucleic Acid Chem., 57: 9.9.1-18.

- Pinheiro et al. (2012) Science, 336: 341-344.

Sesión 2

Genómica y Transcriptómica

Comunicación oral

Chromosome conformation capture assays

Almudena Peláez Ordóñez; Andros Mahiques Mahiques and Mónica Bustos Valdearenas

Genómica Estructural y Funcional

Since the discovery of the double DNA helix until the emergence of -omics techniques there has been a growing interest in discovering how the genes are spatially and temporally distributed and how they interact with each other and with non-coding regions. The answers to these questions started to be answered by chromosome conformation capture techniques. All of these techniques involve capturing the conformation of a segment of the genome or the whole genome at a given time; following fragmentation, ligation of the interacting segments, purification and sequencing. Sequencing methods differentiate different types of capture technologies into: 3C, 4C, 5C and Hi-C. Sequencing strategies are mainly PCR using locus-specific primers (classical 3C), inverse PCR (4C), multiplex sequencing (5C) and tagging ends of restriction fragments with biotinylated nucleotides following sequencing (Hi-C). With these techniques, interaction profiles are obtained that is, how often the physical interaction between the analyzed regions occurs. Techniques 3C and 4C allow obtaining data of concrete loci. They differ in that 3C is only able to compare the interactions of a locus with the surrounding regions whereas 4C is able to compare the same locus with the whole genome (Simonis et al., 2007). 5C and Hi-C are more advanced and are not limited to one locus, offering interaction profiles of multiple regions and generating matrix of interaction frequencies that can cover the entire genome (Dekker et al., 2013). With these techniques, it has been found that even in the same cell population the interaction between the same regions vary and that the DNA undergoes continuous changes in response to stimuli. It is also known that one of the ways in which genomic regions interact is through loops. These technologies have already been used to study genomes such as yeast or human genomes (Dekker et al., 2013), such as regulation of the insulin gene, interaction of the Sonic Hedgehog gene with regions at 1 Mb distance (Williamson et al., 2016) or the importance of non-coding regions in gene regulation (Robyr et al., 2011). Despite all the advantages provided by these technologies, it is necessary to normalize the data since the interaction frequencies are not a real cellular variable. Interaction frequency data can be used to generate 3D conformations of a complete genome and observe how the gene regions are spatially arranged. In spite of the disadvantages that can have, these techniques are essential for the study of the conformation of the genomes.

Bibliography:

-Dekker et al. (2013) Nat Rev Genet, 14: 390-403.

-Robyr et al. (2011) Plos One, 6: e17634.

-Simonis et al. (2007) Nat Methods, 4: 895-901.

-Williamson et al. (2016) Development, 143: 2994-3001.

Panel

Chromatin immunoprecipitation

Juan Antonio Hidalgo Díaz[1] and Sara Fontalva Ostio[2]

[1]Genómica Estructural y Funcional; [2]Fundamentos de Biotecnología Molecular

Chromatin immunoprecipitation (ChIP) is a technique that allows to study the interactions between DNA and proteins2. These interactions are essential for life because they are involved in replication, transcription, regulation, etc. So, it is important to understand how they happen and where.

The first ChIP assay was developed by Gilmour and Lis in the 1980s as a technique for monitoring the association of RNA polymerase II with transcribed and poised genes in Escherichia coli and Drosophila melanogaster2.

This technique has the following steps:

1. Firstly, the cells that express the DNA-binding protein of interest are harvested and treated with a crosslinking agent (formaldehyde, for example) to crosslink proteins to DNA. This process is called Crosslinking4.

2. DNA genome is cut into fragments by sonication or restriction enzymes. Some fragments contain the target protein bound to the DNA1,2.

3. Then specific antibodies, which bind to the specific protein, are used to separate the fragments with this target protein from other fragments that do not have it. The target protein could be alone or forming a complex with other proteins3.

4. Once the fragments are separated, those DNA fragments with the protein are isolated and then this protein is degraded by adding proteinase K1,2,3.

5. Finally, the interesting DNA fragments are purified and ready to be analysed.

To analyze the DNA fragments, there are two different ways: ChIP-on-chip or ChIP-seq2. The use of one of these methods depends on the objective of your study.

The propose of these methods is to find out what sequences our target protein binds to. The ChIP-on-chip1 finds this out by hybridization using a microarray in which there are lots of wells with many specific fragments of the same sequence of the genome. While, ChIP-seq2 does by sequencing these fragments and then they are compared with a known genome.

This technique is useful to detect transcriptional factors, DNA and RNA polymerases, modified histones1 and other proteins which may bind to the DNA. Due to this, it has many applications in epigenetic studies. ChIP is also used to detect differences in genetic expression in two different conditions of a cell or tissue. Besides, it is good to discover where the target protein interacts.

Bibliography:

- Thomas et al. (2009) Methods Mol Biol, 538: 409-423.

- Michael et al. (2009) Cold Spring Harb Protoc, 4: 9.

- Kuo and Allis (1999) Methods, 19: 425-433.

- Sundrud and Rao (2016) Development, 143: 2994-3001.

Panel

MiRNA bank: miRNA isolation and its clinical applications

María Victoria Berlanga Clavero[1]; Juan Antonio García-Sánchez[1]; César Lobato Fernández[2]

[1]Fundamentos de Biotecnología Molecular; [2]Genómica, Proteómica y Metabolómica

MicroRNAs (miRNAs) are a family of small (19-24 nucleotides), non-coding RNAs implicated in gene expression regulation, thus involving in many fundamental cellular processes like cell proliferation, metabolic pathways or immune response (de Planell-Saguer & Rodicio, 2013). They complementary bind to hundreds of 3'UTR regions across the transcriptome, allowing the expression regulation by transcriptional repression or induction of mRNA degradation. Since is known that an abnormal miRNA expression contribute to serious human diseases, identification and annotation of miRNAs have become a major focus (Dweep, Sticht, Pandey, & Gretz, 2011; Femminella, Ferrara, & Rengo, 2015). However, due to its small size and susceptibly to degradation, miRNA isolation and characterization is a huge challenge, so new technologies and data sets are required.

The development of the miRNA bank requires a previous sample isolation. Total RNA is extracted using a standard protocol. Then, purified RNA is run in a 15% PAGE gel obtaining a band which corresponds to 18-30 nucleotides long fragments. The isolated RNA constitutes a set of RNA molecules containing miRNAs.

After being sequenced by RNA-seq, these small fragments are bioinformatically processed. The first step is a phylogenetic comparison. Then, the sequences are searched in the transcriptome. Algorithms obtained by a mobile window on the sequence allow to see the correspondence with the secondary structure of pre-miRNA. A sequence alignment between fragments and mRNA sequences will be performed too.

Throughout the process, sequences won't be discarded and will provide variable values for a final classification using machine learning, which relies on the training set generated from MiRBase (Griffiths-Jones, Grocock, van Dongen, Bateman, & Enright, 2006; Sturm, Hackenberg, Langenberger, & Frishman, 2010).

In conclusion, miRNA bank is a data set of miRNA sequences corresponding to different situations or development stages. Machine learning improves selection and classification of sequences, which is an advantage. Due to their regulation functions, miRNA bank upgrade could be helpful to search therapeutic targets in several diseases like cancer or Alzheimer. In addition, the study of involvement of miRNAs in host-microbiota interaction and host response to bacterial infections may be a good starting point for the development new therapies (Das, Garnica, & Dhandayuthapani, 2016)

Bibliography:

- Das et al. (2016) Frontiers in Cellular and Infection Microbiology, 6: 79.

- de Planell-Saguer & Rodicio (2013) Clinical Biochemistry, 46: 869-878.

- Dweep et al. (2011) Journal of Biomedical Informatics, 44: 839-847.

- Femminella et al. (2015) Frontiers in Physiology, 6: 40.

- Griffiths-Jones et al. (2006) Nucleic Acids Research, 34: D140–D144.

- Sturm et al. (2010) BMC Bioinformatics, 11: 292.

Sesión 3

Proteómica y Metabolómica

Comunicación oral

HiQuant

Miguel Ángel Gallardo Ruiz[1]; Ana Ángel Romero[2] and Francisco Díez de Los Ríos Bravo[2]

[1]Genómica, Proteómica y Metabolómica; [2]Fundamentos de Biotecnología Molecular

HiQuant is a computer program that can be used on Windows and Mac. It uses files generated by other applications as MaxQuant, which is responsible of analyzing proteomes using mass spectrometry. Once we have this type of file, HiQuant is able to open the text file, load it and work with it.

First of all, we have to parse the header of the file, from which we will extract the names of the experiments. We will also select the data type that we have in the file, which may be SILAC or LFQ, and then extract the data contained. Once we have selected the data, we can perform transformations, for example, normalizing the data or creating groups if we have different sub-groups of data in the file.

Finally, we will proceed to prepare the method of analysis of the data, where we will specify the statistical values to be analyzed, the method to be used and the names of the attributes of the experiments, among other characteristics.

Once this is complete, we will generate a series of files (.txt) where all the information will be contained. Using this, we have skiped an arduous task of manual labor and have greatly reduced errors of variation in data and human error. Now, the information can be visualized through HiQuant in a heat map that we can modify according to our needs or you can also generate a file (.xgmml) or (.gexf) to see them in other applications, such as Cytoscape or Gephy, which provide a better visualization of the results obtained.

In our case, we have taken a series of data provided by the official website HiQuant, and we performed an experiment on TK proteins from the MCF-7 cell line and 6,000 other non-TK proteins. These data were analyzed through the configuration provided by HiQuant, obtaining results that we will later analyze with the Cytoscape tool for a better visualization, so we can continue the study of these proteins and their relationship with any type of disease.

Panel

Matrix Assisted Laser Desorption/Ionization-Time of Flight (MALDI-TOF) Mass Spectrometry

Erick Woge Rivera[1], Francisco Bejines López[1] and Trinidad Alba Cano[2]

[1]Genómica Funcional y Estructural; [2]Fundamentos de Biotecnología Molecular

Mass spectrometry (MS) is a powerful spectroscopic tool that allows analyse several types of samples, from elemental to highly complex molecules, through the separation of molecular and atomic species according to their mass. Matrix Assisted Laser Desorption/Ionization Mass Spectrometry (MALDI) is an established technique for analyzing a variety of non-volatile molecules including proteins, peptides, oligonucleotides, lipids, glycans, and other molecules of biological importance (Vestal & Hayden, 2007).

The operation of this tool is initiated by mixing an analyte solution with an organic solution like methanol or aqueous acetone (depends of solvent compatibility with the analyte and strong absorption of radiation at the laser wavelength), then the mix is placed on a stainless-steel sample target and once the solvent is dried, the sample-matrix is irradiated with a high-power laser beam, which simultaneously desorb and ionize the molecules of the sample and the matrix. The packs of ions are accelerated by a fixed electrical potential into the analyzer (flight path) and finally hit the detector. In this way, the mass-to-charge ratio (m/z) values for the analyte ions are calculated as a function of their times of flight, and thus, the mass spectrum is obtained (Monagas et al., 2010). Although sensitivity, selectivity and mass resolution are affected by matrix selection and sample preparation, cationing agents, laser and TOF analyzer, this technique undoubtedly has an enormous potential for studies in Metabolomics and Genomics. For two decades MALDI-TOF MS allowed to determine range in degree of polymerization of the proanthocyanidin (AP) oligomers in apples, characterize the degree of polymerization (DP) and structure of the cranberry PA oligomers and analyze tannins in food and drinks. Nowadays the detection of the linear mode of this technique allows studying proteins, complex mixtures or polymer distributions with high molecular masses (more than 10.000Da), while a reflecting analyzer is required to achieve sufficient resolving power and mass accuracy for analyzing peptides and small molecules. (Vestal & Hayden, 2007).

The versatility of this tool made possible the direct analysis of proteins in biological tissues in the late 80's and in later years allowed the analysis of proteins present in plasma and organelle membranes (receptor proteins, structural proteins, etc.). Recently many studies have focused this technology on the elucidation of molecular differences in cancers, where graded changes in protein distributions from tumour to adjacent normal regions of tissue were shown in clear cell renal cell carcinoma (ccRCC) (Schwamborn & Caprioli, 2010).

Bibliography:

- Monagas et al. (2010) J Pharm Biomed Anal, 51: 358-372.

- Schwamborn and Caprioli (2010) Mol Oncol, 4: 529-538.

- Vestal and Hayden (2007) Int J Mass Spectrom, 268, 83-92.

Sesión 4

Bioinformática

Comunicación oral

Nuclear magnetic resonance, the path towards a recreation of protein structures in solution

José Córdoba Caballero[1], Pablo Rodríguez[2] and José Miguel Valderrama Martín[1]

[1]Fundamentos de Biotecnología Molecular; [2]Genómica Funcional y Estructural

To understand numerous biological processes requires identifying the structure of the molecules. Likewise, knowledge about proteins structures and their domains helps to predict where and how their atoms interact between themselves and other molecules. Hence several techniques and procedures have been developed to approach the configuration and morphology of them. The nuclear magnetic resonance (NMR) is the method that recreates the in vivo states models of proteins. This technique uses the angular moment manifestations of the nuclear spins in certain atoms like 1H. This method, combined with adequate informatics systems, allows us to determine which nuclei get coupled together inside a molecule and make realistic and sensitive approaches about the native structure of polypeptides and their conformational variants. Two types of experiments are performed in order to achieve the proposed goal: homonuclear for sequences with less than 50 amino acids, and heteronuclear for longer polypeptides. Both are interpreted in a similar way, but the latter requires the protein to be isotropically enriched. These procedures generate two dimensional results that grant the ability to determine protons close to each other in the tridimensional structure of the protein, even though they are far apart in the sequence. In other words, first the protein is analyzed purely through different homonuclear and heteronuclear techniques, secondly the nuclear spins systems are sequentially assigned to correlate the nuclei, and finally the structure of the protein is proposed after refining the calculated data a tridimensional (del Río-Portilla, 2003; Kremer & Kalbitzer, 2001). This methodology also provides plenty of important advantages. One of them is the capability to study the structure and interaction between molecules in dissolutions (Bagbyet al., 2001). This has paramount importance because it allows to observe the proteins and the modifications that occur to their conformations while performing their functions in the medium (del Río-Portilla, 2003; Kremer & Kalbitzer, 2001). Moreover, it does not require crystallography, which means a plus since not all proteins can be crystallized and their crystallization may alter the protein structure (Bagby et al., 2001). Last but not least, this technique has an upper hand in comparison to others because it is capable of measuring the molecular movement of the protein within a time scale (Amero, 2011).

Bibliography:

- del Río-Portilla (2003) Mensaje Bioquímico, XXVII: 65-83.

- Kremer and Kalbitzer (2001) Methods Enzymol, 339: 3-19.

- Bagby et al. (2001) Methods Enzymol, 339: 20-41.

- Amero (2011) Mensaje Bioquímico, 35: 159-172.

Panel

Co-expression networks in plants

Ana Álvarez[1], Guillermo López[2] and Francisco Ortigosa[1]

[1]Fundamentos de Biotecnología Molecular; [2]Genómica Funcional y Estructural

Due to transcriptomics techniques development, there exists a wide range of data that must be organised to make its integration and biological interpretation possible. Because of this, gene co-expression networks came up. A gene co-expression network is an undirected graph where each node represents a gene and in which every connection stands for two genes with correlated expression profiles, i.e. correlated expression levels across different conditions.

Gene co-expression network studies allow to visualize graphically the transcriptomics obtained results and make different approximations, such as:

- Explore gene expression levels in different tissues or cell types.

- Perform a differential gene expression analysis.

- Identify which conditions could affect significantly to the expression levels of genes of interest.

Another type of study derived from gene expression networks is the visualization of gene interaction networks and even predict them, which allows us to delve deeper into the importance of the different nodes that make up the real gene network.

There is a huge variety of techniques from which it is possible to obtain co-expression data, e.g. RNA-Seq and exome capture, and there is a large number of databases where it is possible to obtain expression datasets, e.g. GEO, PLEXdb, MGI, etc.

After the data is obtained, it must be preprocessed to make it useful. Once this is done, a data mining method can use the gene expression data to construct a co-expression network.

We present one of the most widely used methods, WGCNA [1]. One of its main advantage is that it creates a weighted network that preserves the scale-free topology, which has been proved to be present in most biological networks [2]. Next, it performs hierarchical clustering to group genes into modules. Enrichment analysis can be used to associate modules with particular functions, which allows us to infer the function of certain genes.

An isolated co-expression network may not be very biologically meaningful, but Big Data provides us with a powerful solution: data integration. Co-expression networks, along with many other sources of data, can be integrated to construct regulatory networks, predict gene functions, reveal the relation between transcriptome and proteome, etc. In plants, cross-species comparison of co-expression networks can be used to enrich networks from not well-known species using well-studied plants, such as Arabidopsis [3]. Several bioinformatics platforms are available for integrating co-expression and functional gene data in plants, such as AraNet, PLANEX, ATTED-II, ComplEX or GeneMANIA.

Bibliography:

- Zhang and Horvath (2005) Stat Appl Genet Mol Biol, 4: Article 17.

- Albert (2005) J Cell Sci, 118: 4947-4957.

- Serin et al. (2016) Front Plant Sci, 7: 444.

www.ingramcontent.com/pod-product-compliance
Lightning Source LLC
Chambersburg PA
CBHW081306170526
45165CB00011B/3439